The Army's Local Economic Effects

2nd Edition

CHRISTOPHER M. SCHNAUBELT, CRAIG A. BOND, COLE SUTERA,
ANTHONY LAWRENCE, JUDITH D. MELE, JOSHUA MENDELSOHN,
CHRISTINA PANIS, MEAGAN L. SMITH

Prepared for the United States Army
Approved for public release; distribution unlimited

For more information on this publication, visit www.rand.org/t/RR1119-1

Library of Congress Cataloging-in-Publication Data is available for this publication.
ISBN: 978-1-9774-0301-8

Published by the RAND Corporation, Santa Monica, Calif.

© 2021 RAND Corporation

RAND® is a registered trademark.

Support RAND
Make a tax-deductible charitable contribution at
www.rand.org/giving/contribute

www.rand.org

Preface

This report documents research and analysis conducted as part of a project entitled *Update on the Army's Local Economic Effects*, sponsored by the Office of the Deputy Chief of Staff, G-8, U.S. Army. The purpose of the project was to provide Army leaders and staff officers with an updated estimate of the economic effects that Army spending produces on U.S. communities and states.

This research was conducted within RAND Arroyo Center's Strategy, Doctrine, and Resources Program. RAND Arroyo Center, part of the RAND Corporation, is a federally funded research and development center (FFRDC) sponsored by the United States Army.

RAND operates under a "Federal-Wide Assurance" (FWA00003425) and complies with the *Code of Federal Regulations for the Protection of Human Subjects Under United States Law* (45 CFR 46), also known as "the Common Rule," as well as with the implementation guidance set forth in U.S. Department of Defense (DoD) Instruction 3216.02. As applicable, this compliance includes reviews and approvals by RAND's Institutional Review Board (the Human Subjects Protection Committee) and by the U.S. Army. The views of sources utilized in this study are solely their own and do not represent the official policy or position of DoD or the U.S. Government.

Contents

Figures and Tables

Figures

Tables

Summary

This report presents findings from RAND Arroyo Center research on the economic activity supported by Army spending at the local level—specifically, within U.S. congressional districts and states. It updates a previous report, *The Army's Local Economic Effects*, by adding tables of data from fiscal years 2015 through 2017.[1]

We estimate the activity supported by Army spending in each of the 435 congressional districts in fiscal years 2014, 2015, 2016, and 2017 using district-level input-output models and a national-level input-output model known as Impact Analysis for Planning.[2] Each district-level model is used to estimate the direct, indirect, and induced effects of Army spending that take place within the district, and these results are used in conjunction with the national-level model to obtain the total economic effects of national-level Army spending on each district and state. Direct effects are calculated as the total Army spending (and associated activity, such as employment) within a district; indirect and induced effects represent the local economic activity that supports both the direct spending and the in-district demand generated from Army spending outside the district. This latter effect often is not calculated with input-output applications. Indirect effects capture interindustry linkages; induced effects capture the effects of household incomes.

This report provides the reader with estimates of the regional-level effects of national-level Army spending, which consists of spending within and outside a region. This regional interpretation is important because the effects are calculated assuming that demand met by suppliers in a given region (in this case, a congressional district or state) is determined outside of that region. In contrast, Army spending at the national level does not meet this assumption because the funding for such spending is derived entirely from taxation of individuals and firms at that level. Although some proportion of national-level Army spending might be funded from a subnational region, this is likely to be small relative to the overall level of demand. Nevertheless, at the regional level, the estimates of economic effects are an upper bound that assumes that the funds for Army expenditures do not come at the expense of decreases in local expenditures

---

[1] Christopher M. Schnaubelt, Craig A. Bond, Frank Camm, Joshua Klimas, Beth E. Lachman, Laurie L. McDonald, Judith D. Mele, Paul Ng, Meagan L. Smith, Cole Sutera, and Christopher Skeels, *The Army's Local Economic Effects*, Santa Monica, Calif.: RAND Corporation, RR-1119-A, 2015a. That edition of this report includes data from only fiscal years 2012 through 2014.

[2] The first edition of this report (Schnaubelt et al., 2015a) uses U.S. Census Bureau data of the 113th Congress for geographic outlines of congressional districts (U.S. Census Bureau, "Cartographic Boundary Shapefiles—Congressional Districts," October 2014). Since this time, the states of Florida, North Carolina, and Virginia have undergone court-mandated redistricting and now use revised congressional districts. There are active lawsuits disputing district lines in several other states, but these had not resulted in any changes at the time of publication. To maintain consistency across the periods of data covered, this report continues to use the geographic district outlines of the 113th Congress for display purposes in the appendixes.

elsewhere. Therefore, estimates of the impact of Army expenditure are more likely to be accu-rate as the difference increases between Army and other federal expenditures coming into a district and the tax revenues leaving that district (i.e., for smaller regions).

Tables S.1 and S.2 summarize the results of the analysis. Table S.1 reports the range of results across the 435 congressional districts of the 113th Congress. All Army direct spending consists of military and government civilian payroll and retiree pay for Regular Army, Army National Guard, and U.S. Army Reserve, plus acquisition and services contracts, by congressional district. Army-driven economic output is the estimate of the value of all produced goods and services in a congressional district that is supported by direct Army spending. All Army personnel and additional employment is a measure of military and civilian personnel directly and indirectly supported by Army spending in congressional districts. Table S.2 reports similar results across states.

Because the model uses Army spending throughout the nation to derive local economic effects, the ratio of economic output to direct spending, often termed the "output multiplier," is likely larger than the economic impact of changes in only local Army spending in a given region for two reasons. First, the model factors in the effects of both in-region and out-of-region Army spending on the economic activity within each district and state, whereas the economic impact analysis would change only final demand within the region. Second, in our analysis, local Army spending in a region is not necessarily equal to the change in final demand in a region as a result of potential subcontracting. Rather, our results are appropriate estimates of the effect of total, nationwide Army spending on each congressional district and state.

Table S.1
Army-Driven Economic Output and All Army Personnel and Additional Employment Congressional District Statistics, 2017 (2016$)

	All Army District Spending ($)	Army-Driven Economic Output ($)	All Army Personnel and Additional Employment
Average	$243.2 million	$882.3 million	8,573
Minimum	$11.2 million	$20.9 million	483
25th percentile	$44.5 million	$143 million	2,290
Median	$80.1 million	$291.9 million	3,801
75th percentile	$195.6 million	$720.5 million	7,344
Maximum	$3.8 billion	$14.6 billion	104,653

Table S.2
Army-Driven Economic Output and All Army Personnel and Additional Employment State Statistics, 2017 (2016$)

	All Army State Spending ($)	Army-Driven Economic Output ($)	All Army Personnel and Additional Employment
Average	$2.1 billion	$7.7 billion	74,590
Minimum	$53.8 million	$210.8 million	1,472
25th percentile	$395.9 million	$1.5 billion	17,435
Median	$1.3 billion	$5.1 billion	48,121
75th percentile	$2.4 billion	$8.5 billion	96,471
Maximum	$11 billion	$42.1 billion	398,390

Acknowledgments

We are grateful for the sponsorship of Timothy Muchmore, who originally initiated this project while serving as the acting director of the Army Quadrennial Defense Review Office. He provided critical support for and oversight of this project in both the current version and the previous edition. RAND reviewers Howard Shatz and James Hosek provided many useful suggestions for improving this report. Programming reviewers Roald Euller and Chuck Stelzner double-checked our coding and ensured that it was properly documented. Josh Russell-Fritch assisted in obtaining updated population, employment, and income data from U.S. Census Bureau surveys.

Abbreviations

ARNG	Army National Guard
DEERS	Defense Enrollment Eligibility Reporting System
DMDC	Defense Manpower Data Center
DoD	U.S. Department of Defense
ESRI	Environmental Systems Research Institute
FPDS-NG	Federal Procurement Data System–Next Generation
FSRS	Federal Subaward Reporting System
FY	fiscal year
I/O	input-output
IMPLAN	Impact Analysis for Planning
NAICS	North American Industry Classification System
RDTE	research, development, test, and evaluation
USAR	U.S. Army Reserve

CHAPTER ONE

Introduction

As a result of the Budget Control Act of 2011 and sequestration, the U.S. Army experienced a budget decrease of approximately $15.35 billion in fiscal year (FY) 2013 and initiated the reduction of more than 106,000 soldiers and civilian employees.[1] Sequestration cuts were not necessary in FYs 2014 and 2015, but Army officials were concerned that if such cuts returned, or if the U.S. Department of Defense (DoD) issued guidance to implement similar reductions regardless of sequestration, the Army's cumulative total budget decreases would have been more than $79 billion from its baseline for FYs 2016–2020.[2]

To help inform decisionmaking if the Army experienced such cuts, the U.S. Army Quadrennial Defense Review Office asked RAND Arroyo Center in May 2014 to provide an empirical understanding of how Army spending affects communities and states to help Army leaders more accurately inform Congress on the distribution of Army personnel and procurement spending and the ripple effects, or *backward linkages*, that it supports. The results of that study covered FYs 2012–2014 and were presented in a three-volume report titled *The Army's Local Economic Effects*.[3]

In September 2017, the Quadrennial Defense Review Office asked RAND Arroyo Center to update the previous publication using data from FYs 2015 and 2016—the most current years for which full data sets were available at the time. While production of the updated report was being finalized, the data set for FY 2017 became available. We therefore added it to the completed analysis.[4] This second edition of *The Army's Local Economic Effects* presents an analysis for FYs 2014 through 2017. The methodology remains the same as the original report, but we have repeated it in the following chapter for the convenience of the reader.

---

[1] U.S. Army Environmental Command, *Supplemental Programmatic Environmental Assessment for Army 2020 Force Structure Realignment*, Washington, D.C., June 2014, p. 1-1.

[2] Interviews with Army G-8 personnel, May 2014.

[3] Christopher M. Schnaubelt, Craig A. Bond, Frank Camm, Joshua Klimas, Beth E. Lachman, Laurie L. McDonald, Judith D. Mele, Paul Ng, Meagan L. Smith, Cole Sutera, and Christopher Skeels, *The Army's Local Economic Effects*, Santa Monica, Calif.: RAND Corporation, RR-1119-A, 2015; Christopher M. Schnaubelt, Craig A. Bond, Frank Camm, Joshua Klimas, Beth E. Lachman, Laurie L. McDonald, Judith D. Mele, Paul Ng, Meagan L. Smith, Cole Sutera, and Christopher Skeels, *The Army's Local Economic Effects: Appendix B, Volume I: Alabama Through Minnesota*, Santa Monica, Calif.: RAND Corporation, RR-1119/1-A, 2015b; Christopher M. Schnaubelt, Craig A. Bond, Frank Camm, Joshua Klimas, Beth E. Lachman, Laurie L. McDonald, Judith D. Mele, Paul Ng, Meagan Smith, Cole Sutera, and Christopher Skeels, *The Army's Local Economic Effects: Appendix B, Volume II: Mississippi Through Wyoming*, Santa Monica, Calif.: RAND Corporation, RR-1119/2-A, 2015c.

[4] We continued to use the 2016 Impact Analysis for Planning (IMPLAN) data set for our input-output (I/O) model rather than purchase the 2017 edition.

How This Report Is Organized

In Chapter Two and Appendix A, we explain the methodology and terminology used in the analysis; in Chapter Three, we describe the sources of data. In Chapter Four, we provide definitions of the key terms used in the district- and state-level reports, we list the summary results across congressional districts and states, and we provide estimates of the economic effects of Army spending at the state level. Appendix B, presented in two separate volumes, contains the detailed results of our analysis organized by state and congressional district.[5]

---

[5]   The first edition of this report (Schnaubelt et al., 2015a) uses U.S. Census Bureau data of the 113th Congress for geographic outlines of congressional districts (U.S. Census Bureau, "Cartographic Boundary Shapefiles—Congressional Districts," October 2014). Since this time, the states of Florida, North Carolina, and Virginia have undergone court-mandated redistricting and now have revised congressional districts. There are active lawsuits disputing district lines in several other states, but these had not resulted in any changes at the time of publication. To maintain consistency across the periods of data covered, this report continues to use the geographic district outlines of the 113th Congress.

CHAPTER TWO

Methodology

This chapter describes the I/O methodology in general, the terminology used, and the specific methodological process used in our analysis.

Input-Output Models

I/O models provide a means of estimating the economic effect of injections of spending (also termed *end-use* or *final demand*) into a regional economy, such as the spending by the Department of the Army in a congressional district. An I/O model is a representation of the linkages between major sectors of a regional economy (and, to a lesser degree, the linkages between these sectors and the rest of the country and rest of the world). Each sector of the regional economy is assumed to require inputs, called *intermediate demand*, from the other sectors to produce output. These inputs can come from local sources (i.e., within the region), from other domestic sources outside the region, or from foreign imports. The total amount of intermediate demand that is sourced from outside the region is called the *leakage* from the local region. The model traces the path of production that satisfies all final demand across industries and sectors, taking into account that a dollar of demand in one sector will stimulate the demand for inputs across other regional sectors, which will subsequently generate additional demands. The model thus calculates the backward linkages of a change in final demand throughout the regional economy, considering the source of the required change in inputs (i.e., taking into account the leakages resulting from the use of out-of-region inputs). According to the structure of the model, the value of inputs used per dollar of output in an industry and the sources of those inputs are unaffected by the change in final demand. Figure 2.1 illustrates these ripple effects through an economy.

I/O model data are organized into tables, or matrices, with each sector of the economy given its own row and column and each region given its own table. Rows identify sales from a specific sector to the other sectors of the economy, identified in the columns. The sum of each row equals the total output for the specified sector in the specified region. Columns contain the inputs used by each sector. They represent the production technology used by an industry in terms of the inputs used from each sector represented in the model. The cells that have the same row and column industry sector name are the intra-industry flows of inputs. I/O models account for profits and wages as inputs, so total sector output equals the total value of sector inputs. In other words, the sum of a row for a sector equals the sum of a column for the same sector. Social accounting matrices, which augment I/O models to decompose final demand into its component parts (household, government, net exports, and investment), can be used

Figure 2.1
Illustration of Backward Linkages in a Regional Economy

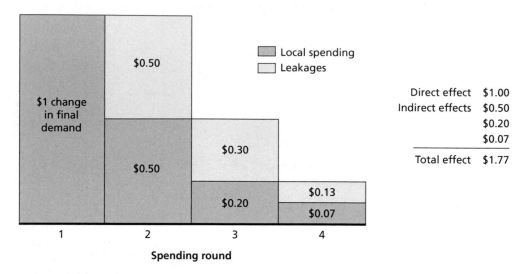

SOURCE: Adapted with permission from Cletus C. Coughlin and Thomas B. Mandelbaum, "A Consumer's Guide to Regional Economic Multipliers," *Federal Reserve Bank of St. Louis Review*, Vol. 73, No. 1, January–February 1991, p. 21, Figure 1.
NOTE: Indirect effects include both intermediate demands generated by industry and incomes generated in region.

to augment the structure and add detail to the models. These matrices form the I/O model known as IMPLAN, available from IMPLAN Group.

For example, consider the (aggregated) manufacturing sector in the national IMPLAN model.[1] The estimated total value of production in this sector for 2012 was approximately $7 trillion (which equals the sum of the manufacturing row and the sum of the manufacturing column). The manufacturing row represents sales from manufacturing to various sectors of the economy. In this case, it shows that this sector sold $88 billion of its output to firms in the agriculture, forestry, fisheries, and hunting sector; $278 billion to firms in the construction sector; and $2 trillion to firms within manufacturing, (i.e., own-sector sales); with the remainder sold to other sectors or for end uses.

The manufacturing column shows the inputs used to produce the $7 trillion manufacturing output. In particular, the sector purchased $216 billion of inputs from the agriculture, forestry, fisheries, and hunting sector; $38 billion from construction; and (consistent with the corresponding row) $2 trillion from firms within manufacturing. The remainder of purchased inputs comes from other industry sectors and labor (i.e., income to households). The column also adjusts for taxes, and it factors in profits to balance the accounts.

The relationships between a change in final demand and overall economic activity in an I/O model are summarized by *multipliers*. Multipliers show the total change in economic activity, given a direct change in final demand or employment. Therefore, they can be interpreted as the total economic change given a one-unit change in final demand.

Indirect effects represent the change given interindustry linkages alone, ignoring the potential effects on household income in the region (termed *open models*). These are referred

---

[1]   This example is taken from the default 11-sector aggregation (from 440 sectors) displayed in the IMPLAN interface.

to as *Type I multipliers*. Potential effects on household income are called *induced effects*, which represent the changes in final demand within a region that occur because of the overall change in labor demand from the direct change. Put differently, induced effects reflect household spending due to wages received resulting from the direct and indirect effects. Multipliers that sum the indirect and induced effects are called *Type II multipliers* and are based on what are known as *closed models*. Because Type II multipliers encompass direct, indirect, and induced effects, they are typically larger than Type I multipliers. Indirect and induced effects are positively related to direct effects, or changes in demand.[2] Table 2.1 provides a reference to the key terms used in the analysis.

Key Assumptions in Input-Output Analysis

I/O models, like all models, make several assumptions regarding the structure of the regional economy and its response to a change in final demand. These assumptions serve to simplify the model and make it tractable. Estimates of observed effects will be different if the

Table 2.1
Key Economic and Mathematical Terms Used in This Report

Term	Definition
End use or final demand	Demand for goods and services that will not subsequently be used in a production process and resold
Intermediate demand	Demand for goods and services that will be used in a production process and ultimately resold to firms
Direct effects	Total Army spending within a district
Indirect effects	Economic activity generated by changes in final demand attributable to interindustry linkages
Induced effects	Economic activity generated by changes in final demand attributable to changes in household incomes
Economic output	Value of all production in an industry or economy-wide
Leakages	Total amount of intermediate demand that is sourced from outside the region
Matrix	A two-dimensional array, or table, of industry sectors with multiple rows and columns
Vector	A one-dimensional array, or table, of industry sectors with multiple rows and one column
Multiplier	Total change in economic activity given a one-dollar change in final demand
Type I multiplier (associated with open model)	Total change in economic activity given a one-dollar change in final demand attributable to interindustry linkages
Type II multiplier (associated with closed model)	Total change in economic activity given a one-dollar change in final demand attributable to interindustry linkages and household incomes

---

[2] The technical term for a change in final demand in this type of analysis is a *shock*. For example, if an industry were to decide to build a new plant in an area, that would be called the initial shock to the system.

assumptions do not hold. Therefore, the assumptions also could be interpreted as limitations. Table 2.2 summarizes these assumptions.

In general, the set of assumptions implicit in I/O analysis results in a lack of flexibility and feedback effects within a regional economy relative to a more complicated model that takes into account price changes and other adaptive behaviors. For example, suppose an I/O model predicts that an increase of $1 million of in-district spending will increase economic output by $1.5 million. The estimate of total economic output is likely a maximum because the spending would tend to stimulate not only intermediate demands but also price changes and other effects. So, the impact might be smaller, especially in the longer run.[3] The advantage to I/O models, however, is the simplicity of the approach. They are especially appropriate for small changes.

Table 2.2
Limitations of the Input-Output Approach

Limitation	Explanation
Fixed production functions and constant returns to scale (linearity)	Inputs for each industry are used in fixed proportions, implying that a doubling of output will require an exact doubling of inputs. No consideration is given to profitability or the potential for substitution between inputs.
Fixed prices	Prices are assumed not to adjust in response to economic factors, and thus firms will not adjust their production on the basis of relative prices. Other macroeconomic feedback and adaptation mechanisms are also excluded.[a]
No supply constraints	Inputs, such as labor, are assumed to be available at prevailing prices. Any constraints that preclude resource availability are not considered.
Constant proportions of local supply	Firms will purchase some fixed portion of their inputs from their local economy and from outside the local economy. The share from outside the local economy determines leakages from the system and is assumed to be constant.
No explicit time dimension	I/O models are static and assume a new equilibrium with a change in final demand.[b]
Perfect mobility of labor	Changes in demand for labor are assumed to be associated with changes in the associated income flows for those workers. An assumed decline in demand results in less economic activity and employment (in fixed proportions by sector), and the wages of the newly unemployed are assumed to leave the region.

SOURCES: Authors' interpretation as informed by Patrick Grady and R. Andrew Muller, "On the Use and Misuse of Input-Output Based Impact Analysis in Evaluation," *Canadian Journal of Program Evaluation*, Vol. 3, No. 2, 1988, pp. 49–61; Coughlin and Mandelbaum, 1991, pp. 19–32; David W. Hughes, "Policy Uses of Economic Multiplier and Impact Analysis," *Choices*, Vol. 18, No. 2, 2nd Quarter, 2003; Rebecca Bess and Zoë O. Ambargis, "Input-Output Models for Impact Analysis: Suggestions for Practitioners Using RIMS II Multipliers," paper presented at the 50th Southern Regional Science Association Conference, New Orleans, La., March 23–27, 2011; and Schnaubelt et al., 2015a.

[a] *Feedback mechanisms* include price changes that result from changes in supply and demand across markets. Other adaptive mechanisms are economic redevelopment efforts, job training, and other policy effects. A type of modeling termed *general equilibrium modeling* accounts for the former and would likely yield different results than those obtained from I/O models.

[b] The length of time needed to establish this new equilibrium is problem- and model-specific.

---

[3]   However, if the increase in Army spending results in additional development activity because of agglomeration (e.g., attracting new firms to the region) or other forward-linkage effects (those that increase demand for firms already in the region), an I/O model might underestimate effects (i.e., estimates would be smaller in magnitude than the actual outcomes).

In addition, any policy responses to a gain or loss in spending are not represented in the I/O structure.[4] One example might be a concerted effort on behalf of planning agencies in the region to develop incentive policies for redevelopment of assets previously owned by the Army. If such efforts are successful, the longer-run negative impacts of a decline in Army spending might be overstated by I/O analysis.

Defining *Local*

We chose the congressional district as the primary unit of analysis for several reasons.[5] The data we used in this study provided for three possibilities at the substate level: zip codes, counties, and congressional districts. Zip codes were immediately deemed impractical because of the sheer number—almost 42,000—and the frequently very small geographic size.[6]

Counties presented a plausible option that we considered during research design. Analysis at the county level would allow the Army to give members of Congress details that congressional district–level analysis would not. However, there are more than 3,100 counties in the United States, compared with 435 congressional districts, which would have made reporting the results more challenging—the report would have been several thousand pages.[7] Furthermore, results at the county level would be less comparable because counties vary significantly in population and economy size. County populations in the United States range from less than 100 in Loving County, Texas, to almost 10,000,000 in downtown Los Angeles County, California.[8] In comparison, congressional district sizes are regulated by law and populations range from approximately 526,000 for Rhode Island's First District to approximately 994,000 for Montana At-Large, although districts' relative structures and economic activity vary significantly.[9]

Using congressional districts as the unit of analysis also should help the Army clearly inform members of Congress regarding the economic effects of Army spending on their constituents—the principal task with which we were asked to assist.

However, using congressional districts has one significant disadvantage—it excludes the District of Columbia and the territories of Guam, the Virgin Islands, and Puerto Rico from the analysis. This means that the economic effects of spending in these areas are not captured by the model; we capture only activity in areas covered by congressional representation. In addition, although IMPLAN estimates regional economic models at the congressional district level (as well as at the county level), the district-level models do not account for trade flows between districts. Our methodology takes this into account, as we will describe.

---

[4] Although these responses are not built into the model structure, they could, in theory, be modeled via additional changes to final demand if such estimates were available or calculated.

[5] As explained previously, this update uses U.S. Census Bureau data of the 113th Congress for geographic outlines of congressional districts as was the case for the first edition of this report (Schnaubelt et al., 2015a; U.S. Census Bureau, 2014).

[6] U.S. Postal Service, *Postal Facts 2014*, p. 19.

[7] U.S. Census Bureau, "USA Counties," webpage, undated-d.

[8] U.S. Census Bureau, "Community Facts," webpage, undated-b.

[9] U.S. Census Bureau, "My Congressional District," webpage, undated-c.

Estimating the Economic Activity Associated with Army Spending at the District Level

To estimate the economic effect of Army spending in each of the 435 congressional districts and the 50 states, we used IMPLAN models at the district level and the national level. Each district-level model was used to estimate the direct, indirect, and induced effects of Army spending taking place within the district. The interindustry relationships were tailored to the unique characteristics of each district using a combination of local and national data as compiled within IMPLAN and summarized using Type II multipliers for each of the 536 sectors contained in the model.[10] The estimate of the effects of any spending activity on a regional economy depended on the structure of the local economy and the size and distribution of the spending itself. Multipliers (or total activity given a change in final demand) varied across regions because of the composition of industries within the economy and the proportion of local inputs used in the production process.

For our analysis, we estimated the regional economic activity associated with all direct Army spending entering the regional and national economies in a fiscal year.[11] We implicitly assumed that this spending originated from outside the region and that direct Army spending in a region represented final demand, unless it was passed through the region as a subcontract. This makes intuitive sense: The purpose of this spending is national defense, and the firms, consumers, and local government within a given region have little control over the ultimate distribution of spending at the national level. However, the district-level models cannot account for the economic activity in the district supported by Army spending in the rest of the country. Thus, Army spending in the rest of the country can generate intermediate demand for outputs from each district that would not be captured by district-specific Army spending and a district-level I/O model.

The national-level IMPLAN model, however, included estimates of industry-specific trade flows across the country and, thus, in principle, captured all the interindustry relationships of national-level Army spending. We used this property to estimate the total effect of national-level Army spending on each district—both final demand generated within the district and the intermediate demands for district products generated by spending outside of the district.[12]

Specifically, we used estimates of the total value of all goods and services produced by the economy (or economic output) as the measure of economic activity. We then apportioned this activity to each congressional district in proportion to each district's share of the sum of district-model effects, which used only the outcomes of direct Army spending within a district. Implicitly, this method modeled all interregional trade flows in proportion to the sector-specific share of national economic output supported by Army spending within a district. This

---

[10] For detailed documentation of IMPLAN models, see IMPLAN Group, homepage, undated-a.

[11] In particular, we adjusted personnel expenditures for federal taxes that do not directly enter the regional (district-level) economy.

[12] Although this procedure is appropriate for apportioning activity to subnational regions, it is not appropriate to use the national-level output figure as an estimate of the national-level impact of Army spending. This is because, at the national level, all government spending is financed through either taxes or borrowing, so that a reduction in Army spending would likely be used to pay down debt, returned to taxpayers, or shifted to other forms of government demand. However, Army spending at the local level can be considered *exogenous*, or determined as outside the economic system rather than inside it, assuming that overall aggregate spending was constant.

assumption of linearity is consistent with the overall structure of I/O models. It also has the advantage of internal consistency, in that the sum of district-level effects can be aggregated to the state level, and the sum of the resultant state-level effects equals the economic activity estimated by the national model and national-level Army spending.[13]

The following procedure details the methods we used to estimate economic output and employment supported by Army spending in the United States.

Step 1: Construct the National and District Models

We used IMPLAN's 536-sector level of detail to create the 435 individual congressional district models and the one national model. In constructing each model, the IMPLAN software created two matrices of 536 rows by 536 columns containing the indirect (Type I) and induced output multipliers, which were summed to obtain Type II multipliers. The interpretation of rows and columns is as described earlier in this chapter. These output multipliers reflect region-specific differences in economic composition and the proportion of inputs that are supplied locally.

Step 2: Estimate Direct Army Expenditures

The inputs to each model were the estimates of sector-specific Army spending in each region, which consisted of spending on active and reserve personnel, civilians, retirees, and procurement and contracting data by place of performance. Army procurement data were sourced from Federal Procurement Data System–Next Generation (FPDS-NG) by place of performance at the zip code level, with industry sectors reported by the appropriate year of the North American Industry Classification System (NAICS) code, the standard industry taxonomy for North American industrial sectors used by federal government statistical agencies.[14] The data were standardized to the 2007 NAICS codes and subsequently converted to the IMPLAN 536-sector classification, as outlined in Appendix A. Deobligations were recorded in the year in which they were recorded in the data.[15] The Army's military and civilian payroll data were sourced from the Defense Manpower Data Center (DMDC) and are reported by home zip code. The spending data at the zip code level were then aggregated to the congressional district level, as outlined in Appendix A. For the military payroll data, all available (pretax) positive compensation categories from the active and reserve pay files were used. These amounts were adjusted for average federal tax withholdings on taxable income by rank using the average rates implied by the DoD Compensation Greenbook for each year.[16] Given the large variation in state and local tax treatment of military pay, we made no adjustments for these withholdings. All else equal, this will slightly overstate take-home pay. Civilian and retiree payroll data reflected actual withholdings for federal, state, and local taxes.

These assignments created 435 district-specific sets of direct Army spending, each with 536 rows (one for each IMPLAN sector) and one column. In mathematical terms and in the

---

[13] The internal consistency property applies to spending flows. Because our primary regional unit is the congressional district and individuals might claim multiple residences over the course of a fiscal year, sums of congressional district personnel counts to state levels or other aggregated levels will likely count some individuals multiple times.

[14] Codes can vary slightly over years.

[15] A deobligation enters as the opposite sign of a planned expenditure. Deobligations have the effect of increasing the variance of direct acquisition and procurement spending from year to year where they occur. Personnel spending is expected to be more constant over time.

[16] DoD, *Military Compensation*, Greenbooks, undated.

technical language of I/O modeling, these sets of numbers are called *vectors* (see Table 2.1). This vector represents goods and services bought by the Army from each of the 536 sectors within the district. Because the original data were in nominal terms, these dollars were converted to 2016 dollars using the gross domestic product deflator provided by the Bureau of Economic Analysis. These vectors represent the final demand that was assumed to be associated with each district. For the national analysis, we summed the 435 district-specific input vectors to create the national analogue.

Step 3: Estimate District-Level Economic Activity from District-Specific Army Spending

Economic effects for each district-level model were calculated by multiplying the Type II output multiplier matrix (536 rows by 536 columns) constructed in Step 1 by the direct-spending vector (536 rows by 1 column) created in Step 2. This produced a table (536 rows by 1 column) representing the overall economic output that direct Army spending supports in the district by sector. These results did not factor in the changes in economic output in each district driven by changes in Army spending in the rest of the country. We applied the same calculations to the national-level model; however, the national region accounted for all relevant trade flows between districts.

At the district level, it is possible that one or more of the industry sectors associated with acquisitions and procurement in the regional IMPLAN model were associated with zero economic activity (i.e., no output or employment for that sector). In these cases, the Type II multiplier is equivalent to zero. Although this is reasonable for subnational regional economies, it does create the potential for a mismatch between reported direct-level Army spending in those sectors and the economic model used to calculate impacts. Potential reasons might be errors in the place of performance or NAICS code of the spending data, errors in the IMPLAN model, or both.

For example, IMPLAN sector 285 is "aircraft engine and engine parts manufacturing" (2007 NAICS code 336412). Although this is a $46 billion industry at the national level, not all district-level models contain positive output and/or employment in this sector (i.e., employment and output in this sector are assumed to be zero for this district). If the procurement and/or acquisitions data used in the analysis identify a positive contract (or negative, in the case of a deobligation) for this sector, either the input data have been misclassified (in the case that true activity in this sector is zero), or the IMPLAN model is incorrect (in the case that true activity is this sector is nonzero).

In these cases, it is assumed that the contracting or subcontracting spending assigned to the zero economic activity sectors is fully passed through the region to other (unspecified) districts. Because this spending is reflected in the national-level spending figures, it is not "lost" but rather reallocated proportionally across districts in accordance with our allocation procedure (see Step 4). We do not believe that this problem will induce major distortions in the analysis, but it is possible that this assumption results in either over- or underestimated overall economic activity in any given district.

Step 4: Adjust for Trade Flows Between Regions

The district-level models potentially underestimate the amount of economic activity in a district supported by total Army spending by ignoring the effects of demand generated by Army spending in the rest of the country. For example, a reduction in personnel spending in Colorado's First Congressional District (CO 1) could affect the demand for certain goods in Virginia's Second

Congressional District (VA 2) if CO 1 imports items from VA 2. However, the district-level models implicitly keep spending in the rest of the country fixed. This implies, for this example, that the reduction in demand in VA 2 is not taken into account in Step 3 of the process.

To account for these effects, we assumed that the national-level economic activity derived from total Army spending and the national I/O model captured all relevant trade flows between regions,[17] and we apportioned these effects (by sector) to each district using that district's share of the sum of district-level effects. We describe this process in the following sections.

Step 4a: Calculate the Sum of the District-Level Effects

To create the denominator of the district-level shares, we summed the 435 district-level vectors of economic output by sector as calculated in Step 3. Because these results exclude trade flows between regions, they will be less than or equal to the national-region estimates.

Step 4b: Calculate District-Level Shares

We calculated the share of national-level effects allocated to each district by dividing each district's output effect from Step 3 by the sum of all districts' output from Step 4a.

Step 4c: Calculate District-Level Effects with Trade Flows

We multiplied the district-level shares from Step 4b by the national output impact in Step 3 by sector to produce 435 district impact vectors adjusted for trade flows. These represented our estimate of the effect of all Army spending nationwide on the economic output of each congressional district.

To summarize:

1. District-level output effects are estimated for each congressional district using direct Army spending in each and the appropriate district-level I/O model.
2. The estimates in (1) are summed across congressional districts and shares for each are computed.
3. National-level output effects are estimated using aggregate Army spending and the national I/O model.
4. Trade-adjusted district-level output effects are estimated by multiplying the national-level effects by the district shares in (2).

By construction, the trade-adjusted district-level effects minus the economic activity supported by direct district-level spending is the trade-flow effect.

In using the national-level model to estimate trade flows, we attempted to capture the impact of the demand generated by direct district-level and indirect out-of-district Army spending on each congressional district. This was necessary because we had no specific information on trade flows between congressional districts and no information on the ultimate place of performance of assumed contracts that are fully passed through to other districts. Despite these limitations, we believe that the methodology used in this report provides a reasonable estimate of the distribution of overall direct and indirect procurement and acquisitions activity across congressional districts using the overall structures of those local economies, and that it provides a means of accommodating conflicts in the data.

---

[17] Implicit in this assumption is that the pattern of trade generated by Army spending is not different from nationwide sector averages. The extent to which this assumption is valid has not, to our knowledge, been formally tested.

Step 5: Estimate Employment Changes

The last task was to estimate the changes in district employment supported by Army spending. We estimated employment directly related to Army personnel (service members and civilian employees) directly from the spending data by counting the number of unique individuals paid by the Army in each region in each fiscal year. These figures included part-time employees.[18]

We estimated employment directly related to procurement and acquisition activity by sector, as well as any employment associated with intermediate and induced demand in a region, using the district-specific ratio of output to employment contained within the IMPLAN software. For all sectors not associated with Army personnel, we multiplied the employment ratio by the sector-specific output estimates from Step 4c to obtain employment estimates.[19] Summing across all sectors in a region yielded the total number of non-Army jobs attributable to Army spending.

Estimating the Economic Activity Associated with Army Spending at the State Level

The state-level estimates of Army-driven economic activity are simply the sum of the district-level effects for each state. This calculation is appropriate because we used the national-level model (and national-level Army spending) to estimate total trade flows across the nation and apportioned these according to district-specific shares from the district-level models. Thus, the calculation implicitly aggregates the intrastate trade flows as represented in the national-level model. As noted, however, the direct Army personnel counts might be overstated if an individual claimed multiple residences in the same state in the same fiscal year. Therefore, the state-level counts will likely overstate overall force strength in a state.

An alternative approach would be to construct the state-level IMPLAN models and use these subregions in the same manner as for the congressional districts (i.e., apportion effects according to Steps 4a through 4c using state data instead of district data). This approach aggregates intrastate (and thus interdistrict for a given state) trade flows via the state-level multipliers associated with each model.

If IMPLAN used estimates of congressional district–level trade flows to aggregate its models, these approaches would be equivalent. However, IMPLAN uses a county-level trade flow model to build multipliers at the multiple county, state, and national levels. So, there might be differences in results at the state level.

We opted to sum the results of the congressional district analyses to obtain state-level estimates to maintain additivity from the district to state to national levels of aggregation.

---

[18] Because we used spending as our input into IMPLAN, we did not use the estimates of direct employment provided by the IMPLAN model in the direct employment calculation.

[19] The government sectors in IMPLAN to which this personnel spending is assigned do not supply outputs to the rest of the economy and have intra-industry multipliers of 1.

Differences in This Analysis and Standard Input-Output Studies

Army Demand from Out of District and Interpretation

A major difference between the methodology used in this study and more-traditional I/O analysis is that our estimate of total economic activity supported by Army spending within a congressional district includes changes in final demand on a district's goods and services from two sources: Army spending directly in the district and Army spending from outside the district that affects demand within the district via supply-chain relationships (or backward linkages). Because the model uses aggregate Army spending (spending in all of the relevant districts/states) to derive local economic effects (effects in one district/state), the ratio of economic output to direct spending, often termed the *output multiplier*, would likely be larger than the economic impact of changes in only local Army spending in a given region. Furthermore, as already discussed, we assumed that any contract or procurement spending assigned to a NAICS code for which there is zero economic output in the appropriate IMPLAN sector at the district level is subcontracted out of the district. The actual change in final demand within a district might be less than that recorded in the direct spending figures, although the total amount of direct Army spending is consistent at the national level. Thus, although the sum of reported district-level direct spending (or, alternatively, the sum of state-level reported direct spending) equals the national direct spending total, the proportion of changes in final demand at the district level might not equal the proportion of reported direct spending at the district level.

As a result, the ratio of total effects to direct Army spending presented in this report might not be an accurate measure of the per-dollar effect of increased local Army spending in a district or state.[20] Rather, our results are appropriate estimates of the effect of total, nationwide Army spending on each congressional district and state.

Because of these issues, these ratios could take on values that are not consistent with single-region I/O analysis. For example, in the case of relatively small demand in a district generated by out-of-district Army spending but a relatively high degree of contract pass-through (thus creating a gap between reported direct spending and the change in final demand), the ratio of total to direct effects might be less than 1. Similarly, in the case of relatively large demand generated by out-of-district Army spending, the ratio of total to direct effects might appear well above 2, even for small geographic areas, such as congressional districts. Therefore, we do not report standard district-level or state-level output and employment multipliers in this report.

In addition, we stress that a regional interpretation of the results is important because the economic effects of nationwide Army spending are calculated assuming that demand met by suppliers in a region (in this case, congressional district or state) is determined outside of that region. In contrast, Army spending at the national level does not meet this assumption because the funding for such spending is derived entirely from taxation of individuals and firms at that level. Although some proportion of national-level Army spending might be funded from a sub-national region, this is likely to be small relative to the overall level of demand. Nevertheless, at the regional level, the estimates of economic effects are an upper bound that assumes that the funds for Army expenditures do not come at the expense of decreases in local expenditures elsewhere. Therefore, the estimates of the impact of Army expenditure are more likely to be

---

[20] Using this report's notation, for example, a standard output multiplier would be calculated by dividing Army-driven economic output by all Army direct spending.

accurate as the difference increases between Army and other federal expenditures coming into a district and the tax revenues leaving a district (i.e., for smaller regions).

Installation-Specific Versus Nationwide District-Level Analysis

Several previous studies have focused on the economic contributions of specific installations on states or regions. For example, Deitrick et al. assesses information in the state of Pennsylvania.[21] In this section, we note a few of the key differences between typical installation-level studies and our work.

Perhaps the biggest difference is in the data that are typically used to drive the I/O models and generate estimated contributions. At the installation level, most "impact" analyses will collect expenditure or employment information related to economic activities at the installation, assign this activity to the appropriate sector in the I/O model, and use the model to generate indirect and induced effects. Although this approach is reasonable to address the economic backward linkages associated with the installation's activities, it does not identify the ultimate *source* of the final demand (i.e., the source of the expenditures) that drives this activity. In our analysis, we restrict attention to final demand generated by (exogenous to the district) Army spending only; specifically, personnel and procurement spending linked to Army-specific expenditures in a particular location (irrespective of installation). In this sense, the scope of the analysis is dissimilar because the assumed direct spending (or, equivalently, final demand) is quite different, naturally leading to differing results. In other words, these alternative approaches might include expenditures associated with installation activities that were not funded directly by the Army.

A second difference is the geographic scope of the analysis. Many local "impact" studies take counties as the smallest geographic unit; for reasons already discussed, we use congressional districts. In general, the smaller the geographic unit, all else equal, the larger the leakages out of the area in terms of indirect or induced spending.

Finally, as noted, our analysis accounts for Army-driven demand for congressional district goods and services from the rest of the country, primarily a result of our focus on the impact of nationwide Army spending. This is not necessary when estimating the contributions of spending at an installation or within a geographic unit regardless of source because all direct spending is restricted to a particular location.

---

[21] Sabina Deitrick, Christopher Briem, Colleen Cain, and Erik R. Pages, *A Comprehensive Assessment of Pennsylvania's Military Installations*, Pittsburgh, Pa.: University of Pittsburgh Center for Social and Urban Research, June 2018.

CHAPTER THREE

Data Sources

This chapter briefly describes the models and data used in analyzing the economic contributions of Army spending. Because the representation of the regional economies in the IMPLAN model (especially at the congressional district level) and the Army spending data are subject to error, estimates of the economic activity associated with Army spending are also subject to error. The relative magnitudes of these errors are unknown (and unknowable), although they are likely smaller at more-aggregated levels of geography. Because pay constitutes roughly 45 percent of the Army budget, this might induce an error of 2–3 percent in our average estimates of Army direct spending. However, the errors are probably not equally distributed, so the error rate might be higher in some congressional districts and states. We have based our estimates on the best available data, and our method of adjusting for trade flows should contribute to error minimization.

The IMPLAN Model

We obtained the regional multipliers and sector spending patterns from the IMPLAN Group. IMPLAN estimates spending patterns using data from the U.S. Bureau of Economic Analysis and the U.S. Census Bureau. It estimates employment numbers from the U.S. Bureau of Labor Statistics and County Business Pattern data. The multipliers are constructed using the Benchmark I/O tables from the U.S. Bureau of Economic Analysis. These models are estimated at the national, state, county, congressional district, and zip code levels.[1] We used the 2016 congressional district models and the national-level model.[2] Trade flows were not estimated at the zip code or congressional district levels, but they were estimated at the county level and up. As described in Chapter Two, Step 4 in our process adjusted our models to estimate the demand of a sector's national output in each district. (See Appendix A for a description of the mapping of zip codes to congressional districts and the mapping of NAICS industry codes to IMPLAN sectors.)

---

[1] IMPLAN Group, "IMPLAN Data Sources," webpage, undated-b.

[2] We note that in using the 2016 model, we are using those relationships as a basis for a particular representation of each district and national model. Subtle or major differences in the structure of the regional and national economies in other years are thus not represented across years. However, absent major regional changes, the effects on the qualitative size of the estimated effects is likely small.

Procurement Data Sources

This report used procurement data for the Department of the Army obtained from FPDS-NG and the Federal Subaward Reporting System (FSRS) via usaspending.gov.[3] The FPDS-NG lists all reported public contract actions made by the government with private contractors (first-tier awardees) valued at $3,000 or more; it excludes contract actions defined as *micro-purchases*, which are made with a government purchase card through the Army government payment card program.[4] It is therefore possible that direct Army spending might be underestimated by as much as 3 percent per year at the national level. FPDS-NG contains obligated dollars per fiscal year by the place of performance (logged by zip code) and the NAICS code. The first-tier contract awards, contracts made directly by the Army, are available at the nine-digit zip code level (zip+4). Zip+4 codes are unique to individual congressional districts. This allows us to identify and apportion all first-tier awardee data to the specific congressional district in which the awarded contracts were performed. Because of the way the data are reported, a multiyear contract might be recorded as completely obligated in one fiscal year. In such cases, our methods would overestimate the spending in that year and underestimate spending in the other years of the contract.

The FSRS contains all contract actions made by first-tier awardees to other private contractors (first-tier subs) for which the contract amount is valued at or above $25,000.[5] We subtracted the amount received by the first-tier sub from the first-tier awardee's place-of-performance zip code and NAICS code total and added the amount to the first-tier sub's place of performance and NAICS codes total. FSRS data are available at the five-digit zip code level. Some zip codes are not unique to congressional districts. For instance, zip code 22407 of Fredericksburg, Virginia, embraces parts of Virginia's First and Seventh Congressional Districts. Per IMPLAN methodology, we divide the economic activity in the zip code region by the number of congressional districts it touches. Additional details are provided in Appendix A.

For state-level tables, we attributed each contract action identified in FSRS to a particular component of the Army by looking at the appropriation used to fund the spending.[6] If the appropriation could be explicitly tied to the ARNG or USAR—for example, "operations and maintenance, Army Guard," or "operations and maintenance, Army Reserve"—we associated the spending with one of those components. If an analogous appropriation—for example, "operations and maintenance, Army"—did not explicitly identify a component, we associated

---

[3]   Specifically, we used all actions initiated by all Army contracting offices.

[4]   In FY 2009, the government payment card program featured approximately 53,300 cards with cardholders making about 4.5 million transactions valued at $4.5 billion. This represented approximately 3 percent of the Army's $140.7 billion-dollar budget request for FY 2009. Inspector General, DoD, *Army Needs to Identify Government Purchase Card High-Risk Transactions*, Washington, D.C., DODIG-2012-043, January 20, 2012; David F. Melcher and Edgar E. Stanton III, "Army FY 2009 Budget Overview," U.S Army briefing, February 2008.

[5]   Office of the Under Secretary of Defense for Acquisition, Technology and Logistics, "Federal Subaward Reporting System (FSRS)," webpage, undated.

[6]   Appropriations in the FPDS-NG data are identified by Treasury Account Symbols for Agency Identifier and Main Account Code. These codes are described in U.S. Treasury, *Federal Account Symbols and Titles (FAST) Book: Agency Identifier Codes*, Washington, D.C., December 2018. However, the Treasury Account Codes necessary to assign Compo (component) and Budget categories became optional as of June 2016. Therefore, we estimated the distribution between components (i.e., Regular Army, Army National Guard [ARNG], and U.S. Army Reserve [USAR]) by using the average proportional shares of each component for each state during FYs 2014–2016 and applying these proportions to Army procurement spending within that state in FY 2017.

it with the Regular Army component. We also associated appropriations relevant only to Regular Army members and families (for example, "family housing construction, Army" and the Homeowners Assistance Fund) with the Regular Army. Finally, we associated all other spending (for example, "procurement or chemical demilitarization construction, defense-wide") with the Regular Army. See Appendix A for more details. This assignment explicitly linked the obligations represented by our data to the type of appropriation that Congress uses to allow the spending. However, it is possible that additional dollars were appropriated to Army components but used for spending not captured by our data and/or not appropriated to the Army specifically, yet the spending was managed by the Army. In this case, we would underestimate the effects of Army spending. Thus, our results do not show the effect of all appropriated dollars to the Army but rather the effect of all obligated dollars contained in FSRS that are associated with the Army.[7]

Personnel Data

Army employment numbers and payroll data were obtained from DMDC and from the Defense Enrollment Eligibility Reporting System (DEERS) for Regular Army, USAR, ARNG, and civilian personnel. DMDC is a military personnel management organization that maintains current and historical records of personnel numbers and of total compensation and benefits received by military personnel.[8] All service members and their eligible family members are required to register with DEERS to receive health care benefits. These records should provide accurate locations for where the majority of payroll dollars accrue and where they are spent.[9]

Although they are not an economic effect per se and their indirect economic effects are accounted for by the model, we also estimate the number of soldiers by component and the number of government civilians employed by the Army in each district. These estimates rely on counting the number of soldiers and government civilians being paid in a particular zip code in September of each year. These numbers change throughout the year as soldiers move (or join or leave service) and we capture the numbers for only one month. However, they should be reasonably reflective of direct employment in a region, except in rare cases for which changes in large numbers of personnel are present, such as base closures. We provided these data at the request of the sponsor because they are an indicator of the Army's physical presence in each district. These data do not strictly reflect an "economic effect," but they are derived from the same sources used to estimate the economic effects and provide useful information for force structure and stationing decisions.[10]

DEERS was used to count soldiers serving in any Army component (Regular Army, USAR, and ARNG) in each domestic zip code in each fiscal year. The DEERS data set is a

---

[7]   Another way to say this is that total spending by the Army contracting offices in the data is not the same as total appropriated spending by the Army according to Treasury Account Symbols. We use all available data for the former but use the codes from the latter to split out this spending by component.

[8]   DMDC, "Welcome to DMDC," webpage, undated.

[9]   The district-level IMPLAN models take into account the average proportion of goods and services purchased locally versus outside of the district.

[10]   During discussions following publication of the original edition of this report (Schnaubelt et al., 2015a), we found that officials were surprised in some cases at the number of soldiers from each component in a district.

monthly file that contains records of each registered service member and his or her dependents. The data were filtered to list only Army personnel, eliminate dependent information, and remove Individual Ready Reserve counts. Unique counts of each service member were then calculated in each fiscal year and aggregated to the district level. Counts do not include service members with missing home zip codes or those with addresses outside of the United States, which could result in an undercount of personnel in some districts.

The active and reserve pay files contain 150 data elements with such information as demographics, special and incentive pays (medical, hazardous duty, bonuses), basic pay, and allowances (e.g., variable housing allowance, overseas housing basic allowance for quarters, federal and state taxes, and separation pay).[11] The civilian pay file consists of extracts from the Defense Civilian Pay System and contains such pay information for DoD civilians as demographics, pay amounts, leave, and hours worked. However, the files available to us do not indicate the components to which the civilian personnel are assigned. Therefore, we use a single category of Government Civilian Personnel that factors in all Army civilians without distinguishing whether they are funded by the Regular Army, USAR, or ARNG.

The active and reserve pay files are monthly individual-level files. The civilian pay files are biweekly. We summed all pay and tax fields at the home zip code level across the months, creating fiscal-year totals. This is a straightforward task after adjusting for missing values coded in the data. The data are summed to the fiscal year and either the five-digit or three-digit zip code level, depending on the source.

Regular Army, USAR, and ARNG payroll data are available at the five-digit zip code level for FYs 2014 through 2017. Total positive compensation for all available fields was used to calculate zip code–level gross pay by rank; average realized federal tax rates on taxable income by rank and fiscal year were calculated from the DoD Compensation Greenbooks.[12] No adjustments were made for state or local taxes. The five-digit zip code totals were apportioned to congressional districts per IMPLAN's methodology (see Appendix A).

Retiree and survivor payroll amounts are available at the three-digit zip code level, as are civilian employment numbers and the counts of retirees and survivors.[13] The retiree and survivor payroll amounts were taken from DoD's *Statistical Report on the Military Retirement System*.[14] The retiree payment amounts are monthly payments made to retired personnel before tax withholdings. Our civilian payroll files contain pre- and post-tax fields. We estimated the amount withheld from retiree payments by dividing the civilian payroll pretax field by the post-tax field and multiplying the result by the retiree payments. To reflect the total payments received each fiscal year by retirees, we multiplied the monthly payment by 12. Retiree payment tax rates are typically lower than tax rates on the employed. Therefore, our calculations likely underestimated the amount received by retired personnel.

---

[11] The data for this project were pulled from the Unit Cohesion File developed and maintained at RAND.

[12] DoD, undated.

[13] See Table A.1 in Appendix A for a description of data available at the nine-, five-, and three-digit zip code levels.

[14] DoD, Office of the Actuary, *Statistical Report on the Military Retirement System: Fiscal Year 2014*, Alexandria, Va., June 2015; DoD, Office of the Actuary, *Statistical Report on the Military Retirement System: Fiscal Year 2015*, Alexandria, Va., July 2016; DoD, Office of the Actuary, *Statistical Report on the Military Retirement System: Fiscal Year 2016*, Alexandria, Va., July 2017.

We converted all spending figures to 2016 dollars using the gross domestic product deflator published by the Bureau of Economic Analysis.[15]

Geographic Data

A geographic information system is a class of software for managing, storing, manipulating, analyzing, visualizing, and using geospatial data. For this study, we used ArcGIS software from Environmental Systems Research Institute. (ESRI). Within this geographic information system, geospatial features are represented as polygons (for larger areas), lines (for linear features), and points (for a point location). The DoD Military Installations, Ranges, and Training Areas data set was used for geographic information on military installation polygons. These installation polygon data are for Regular Army installations, joint installations, and larger-area ARNG sites. Larger-area ARNG polygon sites are installations that have areas larger than 40,000 acres. However, not all of the larger ARNG sites were in this data set, so these sites are represented as points instead. ARNG sites represented as points on the map were provided by the Chief of the Real Estate Branch of the ARNG's Installation Division but were created and curated by ARNG Support for Real Property, R&K Solutions Inc. Geographic information on USAR sites represented as points on the map were provided by the U.S. Army Reserve Command headquarters office. Congressional district data were derived from ESRI and the U.S. Census Bureau's USA 113th Congressional Districts data set.[16] ESRI and U.S. Census Bureau 2013 Populated Places data were used for city points. Base-map data, such as highways, state borders, and water features, were derived from the 2013 ESRI USA Base Map data set. (These data are not pictured in this report.)

---

[15] U.S. Bureau of Economic Analysis, "Table 1.1.9. Implicit Price Deflators for Gross Domestic Product," National Income and Product Accounts Tables, webpage, April 18, 2018.

[16] ESRI, ArcGIS software, USA Base Map data set, 2013; U.S. Census Bureau, 2014; U.S. Census Bureau, "TIGER/Line Shapefiles," April 2015.

CHAPTER FOUR

Results

This chapter provides a brief description of the key terms used in the results at the district and state levels and a summary of the range of those results. We also present results at the state level sorted alphabetically and from highest to lowest effect in terms of economic output. Detailed estimates of the economic activity supported by national Army spending at the congressional district and state levels are provided in Appendix B.[1]

Key Terms and Definitions

The following terms are used in the tables contained in this chapter and in the tables presented in Appendix B.

All Army direct spending is the primary input into the I/O models. It is the estimated amount of Army spending on procurement, personnel (Regular Army, USAR, ARNG, and civilians), and retirees within a region. The figures reported in the tables consist of all reported Army spending within a congressional district, adjusted for outgoing and ingoing subcontracts. As discussed in Chapter Two, however, it is possible that a positive level of spending for a particular NAICS industry in a given district corresponds with an IMPLAN model sector with zero output for that sector. In such cases, we assumed that the contract was passed directly out of the district as a subcontract.[2] This created a gap between all Army direct spending and the change in final demand within a district.

Additional economic output is the estimated amount of economic activity that Army-generated demand supports in a region, as measured by the value of production in all sectors of the economy within a region that is supported as a result of all Army final demand (all Army direct spending minus assumed subcontracting for zero economic activity sectors). This reflects changes in final demand as a result of Army spending within the district and intermediate demand generated from outside the district.

Functionally, this term is calculated as the difference between total economic output supported by Army spending in a region (see *Army-driven economic output*, described next)

---

[1] Appendix B is presented in two separate volumes: Christopher M. Schnaubelt, Craig A. Bond, Cole Sutera, Anthony Lawrence, Judith D. Mele, Joshua Mendelsohn, Christina Panis, and Meagan L. Smith, *The Army's Local Economic Effects: Appendix B, Alabama Through Minnesota*, Santa Monica, Calif.: RAND Corporation, RR-1119/1-1-A, 2021a; and Christopher M. Schnaubelt, Craig A. Bond, Cole Sutera, Anthony Lawrence, Judith D. Mele, Joshua Mendelsohn, Christina Panis, and Meagan L. Smith, *The Army's Local Economic Effects: Appendix B, Mississippi Through Wyoming*, RR-1119/2-1-A, 2021b.

[2] This spending will be captured by the national model estimates, however.

and all Army direct spending. In the case of a high degree of pass-through direct spending or deobligations within a district for a given fiscal year, it is possible for this term to be negative.[3]

For example, assume that total direct Army spending is $500 million in a given district. This spending is assumed to be final demand and will stimulate intermediate demand for inputs used to satisfy this contract within the region. The total value of this supplying activity is additional economic output.

Army-driven economic output is the total value of production in a region that results from all Army direct spending nationwide, including all supplying activity.[4] This measures the estimated direct, indirect, and induced effects of all Army spending within a region and equals the sum of all Army direct spending plus additional economic output. Maintaining the earlier example, if the total intermediate and induced demand results in $250 million of production (additional economic output) from the $500 million of All Army Direct Spending in the district, Army-driven economic output is equal to $250 million + $500 million = $750 million.

All Army direct employment is an estimate of the total number of persons employed by the Army in a region at any time during the fiscal year—Regular Army, USAR, ARNG, and civilians. It is measured as the number of unique individuals paid by the Army in each fiscal year. Because the DMDC data report soldiers in the Active Guard/Reserve program within the active pay files, the number of soldiers assigned to ARNG and USAR units in a district and state was undercounted (all else equal), while Regular Army soldiers were equally overcounted. Dual-status technicians are counted twice because they have two jobs. However, these disparities do not influence the reported estimates of economic effects because spending (not jobs) drives the model. Individuals who have claimed multiple addresses in multiple regions during a fiscal year are counted in each region. Therefore, sums of personnel across regions might not be accurate reflections of the number of individuals employed in a given year. In addition, the counts are of individuals employed at any time during the fiscal year by the Army; thus, these numbers cannot be aggregated to obtain overall force size.

Additional employment is an estimate of the non-Army employment associated with additional economic output, using region-specific ratios of output to employment by sector contained within IMPLAN. It consists of both estimated employment related to direct spending on contracting and procurement and indirect employment generated by the backward linkages in the district-level economy. Continuing with our example, if the average number of jobs per million dollars of output associated with the direct spending for procurement and acquisitions plus generated intermediate demand is 7.3, and contracting and procurement spending was half of total direct spending in the region, the total estimated additional employment would be calculated as 3,650 ([250 + 250] × 7.3).

All Army personnel and additional employment is the total employment supported by Army direct spending nationwide, including all supplying activity. It measures the estimated direct, indirect, and induced jobs supported by all Army spending within a region and is equal to the

---

[3]   In reporting these results, the choices were to (1) report Army direct spending in a district as the actual estimated change in final demand, in which case the sum across all districts would not equal the national-level total of Army direct spending; (2) retain the Army direct spending as reported and calculate additional economic output from the assumed change in final demand, in which case the sum of direct and additional economic output would not equal the total economic output; or (3) calculate additional economic output as the residual, as we did here. We chose the third option as the choice easiest to interpret.

[4]   Output is distinct from value-added or gross regional product. The former provides the total value of all production (including intermediate inputs); the latter excludes the value of intermediate inputs.

sum of all Army direct employment and additional employment. If direct Army employment in the region were equal to 4,750, the total Army personnel and additional employment would be 8,400 (4,750 + 3,650) for the maintained example.

As a reminder, multipliers calculated from the ratio of Army-driven economic output to all Army direct spending should not be used to predict the effects of a change in direct spending because (1) direct spending does not necessarily equal changes in final demand in a district, and (2) the Army-driven economic output figures factor in economic activity generated from both changes in final demand in a district and from intermediate demands generated from Army spending outside the district.

State-Level Tables

In addition to the terms already defined, the state-level tables in Appendix B list economic effects calculated on a component-specific basis. In particular, we assigned direct spending, where possible, to components according to an aggregation of appropriation categories that results in the following categories:

- Regular Army
- USAR
- ARNG
- Civilian/retiree/survivor.

All appropriations categories that could not be mapped explicitly to a component, such as all research, development, test, and evaluation (RDTE) and procurement spending, were assigned to Regular Army. Appendix A provides additional details about the appropriations subcategories that are used in the analysis.

The economic activity associated with each direct spending subcategory was estimated using the methodology described in Chapter Two.

Key Insights and Observations

In this section, we provide a basic summary of the results of the analysis. For more-detailed estimates, see Appendix B. Table 4.1 provides statistics related to total estimated population, employment, and total personal income across the 435 congressional districts for FY 2017. Total personal income as measured by the American Community Survey is a measure of general economic well-being akin to gross district product, which is unavailable at the district level.[5]

Table 4.2 reports statistics about direct Army spending, Army-driven economic output, and Army personnel and additional employment across the 435 congressional districts for FY 2017, reported in 2016 dollars. Again, these effects assume that demand for district-level goods and services as a result of Army spending is generated and predominantly funded through sources outside of each district. Therefore, the results should not be summed to determine national-level effects.

---

[5] *Gross district product* is equal to total district output minus intermediate inputs. It is a measure of the value added by production within a district. Total district-level income is the product of per capita income and population estimates.

Table 4.1
Population, Employment, and Total Personal Income, Congressional District Statistics, 2017

	Total Population	Employed Persons, Population over 16	Total Personal Income (2017$ billions)
Average	747,184	355,585	24.2
Median	745,184	356,411	22.7
Minimum	520,389	199,647	12
25th percentile	717,703	328,567	19.4
75th percentile	772,776	382,597	27.7
Maximum	1,050,493	519,277	64.1

SOURCE: U.S. Census Bureau, "American Community Survey," webpage, undated-a.

NOTES: *Total Personal Income* (calculated from per capita income and population estimates) refers to wage and salary income; net self-employment income; interest; dividends; net rental or royalty income or income from estates and trusts; Social Security or Railroad Retirement income; Supplemental Security Income; public assistance or welfare payments; retirement, survivor, or disability pensions; and all other income. Average, minimum, maximum, and other percentiles are calculated independently for each column.

Table 4.2
Army-Driven Economic Output and All Army Personnel and Additional Employment, Congressional District Statistics, 2017 (2016$)

	All Army District Spending ($)	Army-Driven Economic Output ($)	All Army Personnel and Additional Employment
Average	$243.2 million	$882.3 million	8,573
Median	$80.1 million	$291.9 million	3,801
Minimum	$11.2 million	$20.9 million	483
25th percentile	$44.5 million	$143 million	2,290
75th percentile	$195.6 million	$720.5 million	7,344
Maximum	$3.8 billion	$14.6 billion	104,654

NOTE: Average, minimum, maximum, and other percentiles are calculated independently for each column.

As the table shows, the Army spent approximately $80 million (2016 dollars) in the median district, with a corresponding economic effect of approximately $292 million (2016 dollars) from direct and intermediate demand. This translates into about 3,800 jobs, including all service member and civilian employees of the Army and the private-sector employment supported by Army spending. The range of district-level Army direct spending and effects is large, with the distributions skewed toward a greater number of smaller-impact districts and a small number of high-spending districts. This result is intuitive because the Regular Army personnel and civilian employees account for the greatest proportion of Army spending and would be expected to cluster around installations.

Figure 4.1 provides an additional view of the relationship between district-level Army direct spending and Army-driven economic output in each district. This figure provides further visual evidence of a skewed distribution of direct spending and economic effect, with large spending amounts concentrated in a small number of districts. The straight line in the

Figure 4.1
Scatterplot of Army-Driven Economic Output Against Direct Army Spending, 2017 (2016$ millions)

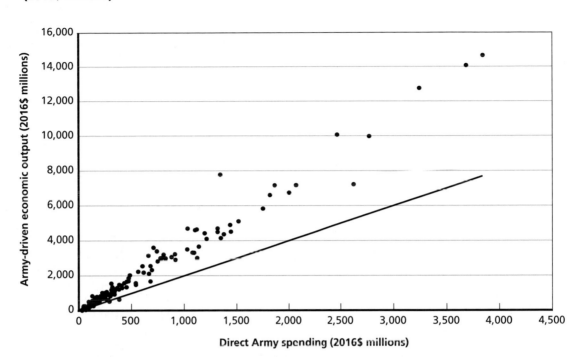

NOTE: The straight line in the figure represents two times the direct district-level Army spending, which is often viewed as an upper bound on output multipliers when considering only direct district spending. This analysis includes demand from all other districts attributable to aggregate Army spending.

figure represents two times the direct district-level Army spending, which is often viewed as an upper bound on output multipliers associated with local spending (as opposed to national-level spending, as is the case here). A linear regression of the data points in the figure shows a slope of 3.65 (*p* value = 0.00), which suggests that every dollar of Army direct spending in a district *plus* the additional demand from the remainder of aggregate Army spending (held constant) results in about $3.65 of economic output in the average district.[6] The inclusion of intermediate demand generated from Army spending outside a district paints a more comprehensive picture of the effect of nationwide Army spending on individual districts.

Table 4.3 provides statistics related to total estimated population, employment, and total personal income across the 50 states for FY 2017.

Table 4.4 presents statistics related to Army-driven economic activity across the 50 states for FY 2017, reported in 2016 dollars.

At the state level, the Army spent approximately $1.3 billion in the median state (in 2016 dollars), with a corresponding economic effect of approximately $5.1 billion, including the effects from direct spending within the state and intermediate demand from outside the state.

---

[6]   The dependent variable is Army-driven economic output (*y*), and the independent variable is direct Army spending (*x*). Estimate with constant term is $y = -8.25 + 3.66x$; *p*-value on constant term is 0.552. Model with no constant is reported $y = 3.65x$.

Table 4.3
Population, Employment, and Total Personal Income, State Statistics, 2017

	Total Population	Employed Persons, Population over 16	Total Personal Income (2017$ billions)
Average	6,500,504	3,093,591	210.3
Median	4,569,261	2,007,833	130.7
Minimum	579,315	282,202	17.9
25th percentile	1,815,857	786,913	51.3
75th percentile	7,405,743	3,599,753	286.9
Maximum	39,536,653	18,757,501	1,385.6

SOURCE: U.S. Census Bureau, undated-a.

NOTES: *Total Personal Income* (calculated from per-capita income and population estimates) refers to wage and salary income; net self-employment income; interest; dividends; net rental or royalty income or income from estates and trusts; Social Security or Railroad Retirement income; Supplemental Security Income; public assistance or welfare payments; retirement, survivor, or disability pensions; and all other income. Average, minimum, maximum, and other percentiles are calculated independently for each column.

Table 4.4
Army-Driven Economic Output and All Army Personnel and Additional Employment, State Statistics, 2017 (2016$)

	All Army District Spending ($)	Army-Driven Economic Output ($)	All Army Personnel and Additional Employment
Average	$2.1 billion	$7.7 billion	74,590
Median	$1.3 billion	$5.1 billion	48,121
Minimum	$53.8 million	$210.8 million	1,472
25th percentile	$395.9 million	$1.5 billion	17,435
75th percentile	$2.4 billion	$8.5 billion	96,471
Maximum	$11.0 billion	$42.1 billion	398,390

NOTE: Average, minimum, maximum, and other percentiles are calculated independently for each column.

This translates to slightly more than 48,000 Army and non-Army jobs. The range of state-level Army direct spending and effect is not quite as large, proportionally, as it is in congressional districts, with more symmetric (though still skewed) distributions.

Figure 4.2 shows the cumulative percentage of the economic activity supported by all Army spending across states for FY 2017 (in 2016 dollars), sorted from highest to lowest effect by state, from left to right.

The figure confirms the skew of the total impact distribution, with the top five states (Virginia, Texas, California, Florida, and Alabama) accounting for 37 percent of total Army-driven economic output, and the bottom five states (Idaho, Montana, Maine, Rhode Island, and Wyoming) constituting less than one-half of 1 percent of total Army-driven economic output across all states.

Figure 4.2
Cumulative Percentage of Economic Activity Supported by Army Spending by State, Army-Driven Economic Output and All Army Personnel and Employment, FY 2017 (2016$)

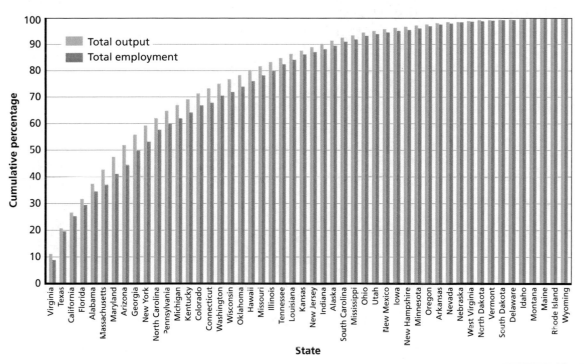

NOTE: States are ordered from largest to smallest Army-driven economic output. Because this is a cumulative figure, each column builds on the one before it. Therefore, states with the least output have the longest bars.

Figure 4.3
Army-Driven Economic Output Per Capita (2016$)

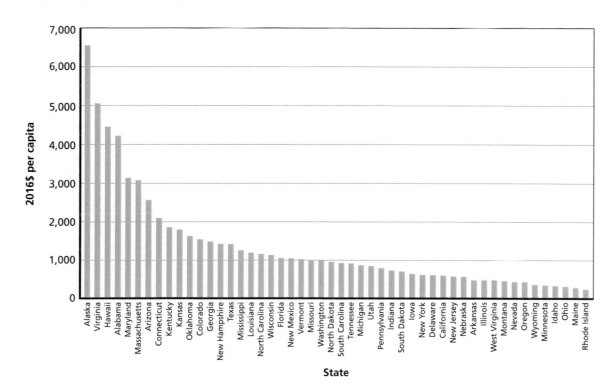

States have a much larger variation in economic activity and population than congressional districts do; therefore, we present the Army-driven economic output per capita in Figure 4.3. The range is quite large, with the greatest per capita effect in Alaska, at just over $6,500 per person (2016 dollars), and the smallest effect in Rhode Island, at $228 per person. Once again, the results are skewed, with a mean per capita effect of $1,343 and a median (South Carolina and North Dakota) of $946 per person.

Finally, Tables 4.5 and 4.6 present the results of the economic effects of total Army spending by state for FY 2017. Table 4.5 presents results alphabetically; Table 4.6 presents results sorted from largest to smallest economic impact in terms of total Army-driven economic output.

Table 4.5
Economic Effects of Total Army Spending by State, Alphabetically Sorted, 2017 (2016$)

State	All Army State Spending[a] ($)	Additional Economic Output[b] ($)	Army-Driven Economic Output[c] ($)	Army-Driven Economic Output per Capita ($)	All Army Direct Employment[d]	Additional Employment[e]	All Army Personnel and Additional Employment[f]
Alabama	5,388,431,279	15,201,420,781	20,589,852,060	4,217	51,128	129,888	181,016
Alaska	1,312,891,238	3,413,423,554	4,726,314,792	6,548	22,717	24,826	47,543
Arizona	5,740,917,152	11,447,227,522	17,188,144,674	2,573	30,257	85,512	115,769
Arkansas	365,632,533	1,065,359,033	1,430,991,566	478	15,617	8,487	24,104
California	6,749,631,686	15,721,072,648	22,470,704,334	587	101,462	106,496	207,958
Colorado	2,232,184,580	5,840,719,913	8,072,904,493	1,531	61,055	46,651	107,706
Connecticut	2,090,320,203	5,433,291,136	7,523,611,339	2,094	741	28,344	29,085
Delaware	166,748,311	405,842,892	572,591,203	609	3,318	2,989	6,307
Florida	5,706,719,322	14,938,254,137	20,644,973,459	1,056	50,594	111,337	161,931
Georgia	4,069,975,476	10,838,520,474	14,908,495,950	1,488	109,258	95,200	204,458
Hawaii	1,980,681,704	4,254,604,721	6,235,286,425	4,460	45,126	32,925	78,051
Idaho	124,286,707	431,624,940	555,911,647	340	6,901	3,075	9,976
Illinois	1,757,813,047	4,365,459,999	6,123,273,046	475	31,605	32,492	64,097
Indiana	1,271,680,089	3,482,865,237	4,754,545,326	722	20,735	19,663	40,398
Iowa	605,286,263	1,340,081,139	1,945,367,402	626	15,045	9,300	24,345
Kansas	1,311,724,070	3,918,684,723	5,230,408,793	1,794	45,185	30,203	75,388
Kentucky	1,953,182,057	6,251,523,739	8,204,705,796	1,866	29,794	53,393	83,187
Louisiana	1,305,409,818	4,240,990,592	5,546,400,410	1,185	30,917	36,041	66,958
Maine	110,139,094	266,244,661	376,383,755	280	392	2,008	2,400
Maryland	5,000,580,656	13,605,769,833	18,606,350,489	3,140	48,688	107,129	155,817

Table 4.5—Continued

State	All Army State Spending[a] ($)	Additional Economic Output[b] ($)	Army-Driven Economic Output[c] ($)	Army-Driven Economic Output per Capita ($)	All Army Direct Employment[d]	Additional Employment[e]	All Army Personnel and Additional Employment[f]
Massachusetts	4,979,041,899	15,522,941,089	20,501,982,988	3,081	2,671	94,998	97,669
Michigan	2,455,173,627	6,138,962,157	8,594,135,784	873	23,994	40,968	64,962
Minnesota	653,416,642	1,222,735,683	1,876,152,325	346	17,846	8,525	26,371
Mississippi	1,065,977,957	2,689,961,291	3,755,939,248	1,249	18,172	23,957	42,129
Missouri	1,608,907,905	4,587,685,018	6,196,592,923	1,021	46,734	36,319	83,053
Montana	103,196,500	353,198,967	456,395,467	448	4,689	2,872	7,561
Nebraska	255,019,108	799,971,798	1,054,990,906	560	8,923	6,012	14,935
Nevada	377,002,548	844,465,571	1,221,468,119	428	9,178	6,599	15,777
New Hampshire	494,207,206	1,424,289,826	1,918,497,032	1,436	695	10,036	10,731
New Jersey	1,281,339,215	3,741,481,112	5,022,820,327	564	6,211	28,274	34,485
New Mexico	605,075,330	1,582,977,288	2,188,052,618	1,055	8,501	13,911	22,412
New York	3,367,553,759	8,908,253,822	12,275,807,581	625	62,312	59,871	122,183
North Carolina	3,224,579,049	8,517,912,733	11,742,491,782	1,179	100,266	67,905	168,171
North Dakota	180,765,767	528,445,682	709,211,449	952	4,851	3,759	8,610
Ohio	1,078,952,269	2,537,636,653	3,616,588,922	313	26,882	17,949	44,831
Oklahoma	1,593,028,493	4,755,539,141	6,348,567,634	1,629	33,921	38,744	72,665
Oregon	452,503,465	1,232,584,288	1,685,087,753	427	13,245	9,822	23,067
Pennsylvania	3,117,984,584	6,982,797,111	10,100,781,695	793	44,661	48,217	92,878
Rhode Island	93,489,025	147,068,769	240,557,794	228	374	1,098	1,472
South Carolina	1,201,607,230	3,315,890,142	4,517,497,372	939	29,910	24,091	54,001

Table 4.5—Continued

State	All Army State Spending[a] ($)	Additional Economic Output[b] ($)	Army-Driven Economic Output[c] ($)	Army-Driven Economic Output per Capita ($)	All Army Direct Employment[d]	Additional Employment[e]	All Army Personnel and Additional Employment[f]
South Dakota	139,192,605	465,926,026	605,118,631	713	5,497	3,538	9,035
Tennessee	1,805,869,841	4,277,934,515	6,083,804,356	924	61,611	30,235	91,846
Texas	10,093,489,727	27,520,100,790	37,613,590,517	1,402	198,738	199,652	398,390
Utah	655,280,399	1,810,336,866	2,465,617,265	845	12,637	12,353	24,990
Vermont	372,389,104	280,110,420	652,499,524	1,034	419	3,767	4,186
Virginia	10,989,202,629	31,105,597,450	42,094,800,079	5,076	90,675	241,812	332,487
Washington	2,163,327,744	4,787,217,381	6,950,545,125	992	72,401	36,001	108,402
West Virginia	268,392,435	590,693,481	859,085,916	466	7,388	5,399	12,787
Wisconsin	1,831,608,051	4,712,726,932	6,544,334,983	1,144	18,190	30,509	48,699
Wyoming	53,846,207	156,954,009	210,800,216	363	3,150	1,113	4,263

[a] Consists of military and government civilian payroll and retiree pay for Regular Army, ARNG, and USAR plus acquisition and services contracts in district. Direct spending in sectors with zero output in IMPLAN district model is assumed to be passed out of district. Does not include demand from Army spending from the rest of the country.

[b] Consists of estimated indirect and induced effects of all Army direct spending within and outside of district.

[c] All Army direct spending assumed to stay in district plus estimated additional economic output. Does not include spending passed out of district. Might not sum exactly due to rounding.

[d] Consists of military and government civilian personnel for Regular Army, ARNG, and USAR, including soldiers not on active duty. Sum of unique individuals paid in each congressional district by state in each fiscal year.

[e] Estimated additional jobs resulting from additional economic output, including indirect and induced effects of all Army direct spending within and outside of district.

[f] All Army employment plus estimated additional employment. Might not sum exactly due to rounding.

Table 4.6
Economic Effects of Total Army Spending by State, Sorted Largest to Smallest by Army-Driven Economic Output, 2017 (2016$)

State	All Army State Spending[a] ($)	Additional Economic Output[b] ($)	Army-Driven Economic Output[c] ($)	Army-Driven Economic Output per Capita ($)	All Army Direct Employment[d]	Additional Employment[e]	All Army Personnel and Additional Employment[f]
Virginia	10,989,202,629	31,105,597,450	42,094,800,079	5,076	90,675	241,812	332,487
Texas	10,093,489,727	27,520,100,790	37,613,590,517	1,402	198,738	199,652	398,390
California	6,749,631,686	15,721,072,648	22,470,704,334	587	101,462	106,496	207,958
Florida	5,706,719,322	14,938,254,137	20,644,973,459	1,056	50,594	111,337	161,931
Alabama	5,388,431,279	15,201,420,781	20,589,852,060	4,217	51,128	129,888	181,016
Massachusetts	4,979,041,899	15,522,941,089	20,501,982,988	3,081	2,671	94,998	97,669
Maryland	5,000,580,656	13,605,769,833	18,606,350,489	3,140	48,688	107,129	155,817
Arizona	5,740,917,152	11,447,227,522	17,188,144,674	2,573	30,257	85,512	115,769
Georgia	4,069,975,476	10,838,520,474	14,908,495,950	1,488	109,258	95,200	204,458
New York	3,367,553,759	8,908,253,822	12,275,807,581	625	62,312	59,871	122,183
North Carolina	3,224,579,049	8,517,912,733	11,742,491,782	1,179	100,266	67,905	168,171
Pennsylvania	3,117,984,584	6,982,797,111	10,100,781,695	793	44,661	48,217	92,878
Michigan	2,455,173,627	6,138,962,157	8,594,135,784	873	23,994	40,968	64,962
Kentucky	1,953,182,057	6,251,523,739	8,204,705,796	1,866	29,794	53,393	83,187
Colorado	2,232,184,580	5,840,719,913	8,072,904,493	1,531	61,055	46,651	107,706
Connecticut	2,090,320,203	5,433,291,136	7,523,611,339	2,094	741	28,344	29,085
Washington	2,163,327,744	4,787,217,381	6,950,545,125	992	72,401	36,001	108,402
Wisconsin	1,831,608,051	4,712,726,932	6,544,334,983	1,144	18,190	30,509	48,699
Oklahoma	1,593,028,493	4,755,539,141	6,348,567,634	1,629	33,921	38,744	72,665
Hawaii	1,980,681,704	4,254,604,721	6,235,286,425	4,460	45,126	32,925	78,051

Table 4.6—Continued

State	All Army State Spending[a] ($)	Additional Economic Output[b] ($)	Army-Driven Economic Output[c] ($)	Army-Driven Economic Output per Capita ($)	All Army Direct Employment[d]	Additional Employment[e]	All Army Personnel and Additional Employment[f]
Missouri	1,608,907,905	4,587,685,018	6,196,592,923	1,021	46,734	36,319	83,053
Illinois	1,757,813,047	4,365,459,999	6,123,273,046	475	31,605	32,492	64,097
Tennessee	1,805,869,841	4,277,934,515	6,083,804,356	924	61,611	30,235	91,846
Louisiana	1,305,409,818	4,240,990,592	5,546,400,410	1,185	30,917	36,041	66,958
Kansas	1,311,724,070	3,918,684,723	5,230,408,793	1,794	45,185	30,203	75,388
New Jersey	1,281,339,215	3,741,481,112	5,022,820,327	564	6,211	28,274	34,485
Indiana	1,271,680,089	3,482,865,237	4,754,545,326	722	20,735	19,663	40,398
Alaska	1,312,891,238	3,413,423,554	4,726,314,792	6,548	22,717	24,826	47,543
South Carolina	1,201,607,230	3,315,890,142	4,517,497,372	939	29,910	24,091	54,001
Mississippi	1,065,977,957	2,689,961,291	3,755,939,248	1,249	18,172	23,957	42,129
Ohio	1,078,952,269	2,537,636,653	3,616,588,922	313	26,882	17,949	44,831
Utah	655,280,399	1,810,336,866	2,465,617,265	845	12,637	12,353	24,990
New Mexico	605,075,330	1,582,977,288	2,188,052,618	1,055	8,501	13,911	22,412
Iowa	605,286,263	1,340,081,139	1,945,367,402	626	15,045	9,300	24,345
New Hampshire	494,207,206	1,424,289,826	1,918,497,032	1,436	695	10,036	10,731
Minnesota	653,416,642	1,222,735,683	1,876,152,325	346	17,846	8,525	26,371
Oregon	452,503,465	1,232,584,288	1,685,087,753	427	13,245	9,822	23,067
Arkansas	365,632,533	1,065,359,033	1,430,991,566	478	15,617	8,487	24,104
Nevada	377,002,548	844,465,571	1,221,468,119	428	9,178	6,599	15,777
Nebraska	255,019,108	799,971,798	1,054,990,906	560	8,923	6,012	14,935

Table 4.6—Continued

State	All Army State Spending[a] ($)	Additional Economic Output[b] ($)	Army-Driven Economic Output[c] ($)	Army-Driven Economic Output per Capita ($)	All Army Direct Employment[d]	Additional Employment[e]	All Army Personnel and Additional Employment[f]
West Virginia	268,392,435	590,693,481	859,085,916	466	7,388	5,399	12,787
North Dakota	180,765,767	528,445,682	709,211,449	952	4,851	3,759	8,610
Vermont	372,389,104	280,110,420	652,499,524	1,034	419	3,767	4,186
South Dakota	139,192,605	465,926,026	605,118,631	713	5,497	3,538	9,035
Delaware	166,748,311	405,842,892	572,591,203	609	3,318	2,989	6,307
Idaho	124,286,707	431,624,940	555,911,647	340	6,901	3,075	9,976
Montana	103,196,500	353,198,967	456,395,467	448	4,689	2,872	7,561
Maine	110,139,094	266,244,661	376,383,755	280	392	2,008	2,400
Rhode Island	93,489,025	147,068,769	240,557,794	228	374	1,098	1,472
Wyoming	53,846,207	156,954,009	210,800,216	363	3,150	1,113	4,263

[a] Consists of military and government civilian payroll and retiree pay for Regular Army, ARNG, and USAR plus acquisition and services contracts in district. Direct spending in sectors with zero output in IMPLAN district model is assumed to be passed out of district. Does not include demand from Army spending from the rest of the country.

[b] Consists of estimated indirect and induced effects of all Army direct spending within and outside of district.

[c] All Army direct spending assumed to stay in district plus estimated additional economic output. Does not include spending passed out of district. Might not sum exactly due to rounding.

[d] Consists of military and government civilian personnel for Regular Army, ARNG, and USAR, including soldiers not on active duty. Sum of unique individuals paid in each congressional district by state in each fiscal year.

[e] Estimated additional jobs resulting from additional economic output, including indirect and induced effects of all Army direct spending within and outside of district.

[f] All Army employment plus estimated additional employment. Might not sum exactly due to rounding.

CHAPTER FIVE

Conclusion

This report presents findings on the economic activity supported by Army spending on state and local economies. Using a combination of congressional district and national-level I/O models, in conjunction with procurement and payroll data, we estimated the regional economic activity associated with Army-generated demand. Given the lack of feedback associated with the I/O methodology, the estimates should be interpreted as the upper bounds of economic activity associated with Army spending at the local level.

In addition, because we estimated the effects of both in-region and out-of-region total Army spending on the economic activity within each district and state, the results reported in the district- and state-level tables should not be used to calculate the per-dollar effect of increased or decreased Army spending in a district or state. Rather, for any given suite of cuts or spending increases that can be associated with a geographic area, the methodologies detailed in this report could be used to estimate effects, but per-dollar results would likely vary because of differences in the distribution of demand changes across local and nonlocal sectors and the geographic distribution of the suite of cuts. In addition, only net demand changes should be used; that is, any spending changes by the Army should be offset by any spending changes made by other agencies or sources of exogenous demand as a result of decreased Army demand.

We found that the Army directly spent approximately $80.1 million in the median district and $1.3 billion in the median state in FY 2017, with considerable variance across the local economies. This direct spending and the intermediate demands generated by out-of-district and out-of-state spending contributed a total of $291.9 million of economic output to the median district and $5.1 billion to the median state.[1] This translates into about 3,801 jobs for the median district and more than 48,000 for the median state, with a wide range across economies.

---

[1] All figures converted to 2016 dollars.

Preprocessing of Direct Army Spending Data

Mapping Zip Code Data to Congressional Districts

Table A.1 summarizes the data we used in our economic impact models and the length of the corresponding zip code.

Five-Digit Zip Code to Congressional District

Out of 34,902 zip codes, 6,310 are in at least two districts. In these instances, IMPLAN apportions the economic activity in the zip code according to how many congressional district boundaries the zip code crosses; if the zip code lies in three congressional districts, each district is apportioned one-third of the zip code's economic activity. Also, there were 932 five-digit zip codes in the Regular Army, USAR, and ARNG pay files that did not map to congressional districts. DMDC did note that foreign postal codes are sometimes entered into the zip field. For instance, 06686 is a German postal code. Another possibility is that those 932 zip codes were erroneously entered by the user. If this is the case, the pay in these zip codes should be apportioned to the best alternative zip code that does exist within the United States. Additionally, some zip codes are in the United States but do not belong to a congressional district. Zip code 80279, the Air Force Accounting and Finance Center in Denver, had a total pay of $14 million in FY 2013. Considering the two latter circumstances, we truncated the zip codes to three-digit numbers and apportioned the pay as outlined in the next section.

Table A.1
Model Data and Associated Zip Code Length

Data	Length of Zip Code (Digits)
First-tier awards (FPDS-NG)	9
First-tier subawards (FSRS)	5
Regular Army payroll and personnel counts	5
USAR payroll and personnel counts	5
ARNG payroll and personnel counts	5
Civilian payroll and personnel counts	5[a]
Retiree and survivor pay and counts	3

[a] As of FY 2017, DMDC provides only three-digit zip codes for civilian payroll and personnel counts.

Three-Digit Zip Code to Congressional District

The following steps were taken to apportion the three-digit zip code payroll and employment counts to congressional districts:

1. Calculate the total population in the truncated three-digit zip code from the 2016 ESRI U.S. Zip Code Points Geodatabase.[1]
2. Divide the five-digit zip code point population estimate by the three-digit zip code point total. This gives the percentage of the total population of the three-digit zip code living in the five-digit zip code.
3. For each unique five-digit zip code in our crosswalk of zip code to congressional district, multiply the percentage in step 2 by the three-digit totals in our DMDC files, and apportion that amount (either payroll amounts or employee counts) to the five-digit zip code.
4. Apportion the five-digit zip code data from step 3 to congressional districts per IMPLAN's methodology, as described in the Five-Digit Zip Code to Congressional District section of this appendix.

Mapping NAICS Industries to IMPLAN Sectors

Since 2015, the year we wrote *The Army's Local Economic Effects*, IMPLAN changed its sector designation from a 440-sector classification to a 536-sector classification.[2] IMPLAN provides bridges from NAICS codes into the IMPLAN 536-sector classification, and a bridge from the 440 to 536 sectors is publicly available from the IMPLAN website.

Consequently, some sectors contain multiple NAICS codes. The IMPLAN models impose the same regional multiplier on all NAICS codes contained in the same IMPLAN sector. IMPLAN provides a crosswalk from NAICS codes to the appropriate sector in the IMPLAN 536-sector classification. However, IMPLAN does not provide a crosswalk between construction NAICS and IMPLAN 536-sector classification. IMPLAN does have a crosswalk between the 2007 construction NAICS and the 440-sector classification; it also has a crosswalk between the 440-sector classification and the 536-sector classification. We used this combination of crosswalks to map our procurement and payroll data to the IMPLAN sectors.

There were seven congressional districts in which 68–95 percent of direct Army spending occurred in sectors with multipliers equal to zero. In these instances, we chose the best alternative sector that had a nonzero multiplier in the district based on whether there was spending in the alternate sector and how similar the alternate sector was to the original sector. For instance, in Indiana's 2nd Congressional District in FY 2016, 95 percent of total Army direct spending in the district ($661.2 million) occurred in IMPLAN sectors with a multiplier of zero. Of that, $658.6 million was apportioned to IMPLAN Sector 366 (military armored vehicle, tank, and tank component manufacturing). IMPLAN Sector 367 (all other transportation equipment manufacturing) had a nonzero multiplier and had positive direct Army spending. Spending from Sector 366 was redirected to Sector 367.

---

[1] ESRI, ArcGIS software, U.S. ZIP Code Points File Geodatabase Feature Class, 2016.

[2] IMPLAN Group, "Release Notes," webpage, undated-c.

The NAICS codes in the FPDS-NG include 1997, 2002, 2007, and 2012 NAICS. We crosswalked the FPDS-NG data attached to the 1997, 2002, and 2012 NAICS into the 2007 NAICS and then crosswalked the 2007 NAICS to the IMPLAN 440-sector classification. Finally, we crosswalked the 440-sector classification to the 536-sector classification.

Component Direct Spending Categories Based on Appropriation Category

Regular Army consists of the following appropriations categories:

- Military Personnel, Army
- Medicare-Eligible Retiree Health Fund Contribution, Army
- Operation and Maintenance, Army
- Military Construction, Army
- Family Housing Construction, Army
- Family Housing Operation and Maintenance, Army
- Homeowners Assistance Fund
- Aircraft Procurement, Army
- Missile Procurement, Army
- Procurement of Weapons and Tracked Combat Vehicles, Army
- Procurement of Ammunition, Army
- Other Procurement, Army
- Joint Improvised Explosive Device Defeat Fund (A Portion)
- Research, Development, Test and Evaluation, Army
- Chemical Agents and Munitions Destruction, Defense
- Chemical Demilitarization Construction, Defense-Wide
- Working Capital Fund, Army
- Environmental Restoration, Army
- Environmental Restoration, Formerly Used Defense Sites
- U. S. Army National Cemeteries Program
- Afghanistan Security Forces Fund
- Corps of Engineers—Civil Works
- Other Defense—Civil Programs.

USAR consists of the following appropriations categories:

- Reserve Personnel, Army
- Medicare-Eligible Retiree Health Fund Contribution, Reserve Personnel, Army
- Operation and Maintenance, Army Reserve
- National Guard and Reserve Equipment (A Portion)
- Military Construction, Army Reserve.

ARNG consists of the following appropriations categories:

- National Guard Personnel, Army
- Medicare-Eligible Retiree Health Fund Contribution, National Guard Personnel, Army
- Operation and Maintenance, Army National Guard
- National Guard and Reserve Equipment (A Portion)
- Military Construction, Army National Guard.

We classified Regular Army direct spending as consisting of Regular Army military personnel, operations and maintenance, military construction, procurement, all RDTE, and spending not explicitly associated with other components. In FY 2014, Army procurement accounted for about 11 percent of the Regular Army direct spending, although both the Reserve and Guard also benefit from Army procurement. About 60 percent of Army procurement benefits the Regular Army, 10 percent benefits the Reserve, and 30 percent benefits the Guard based on their proportionate shares of capital/materiel as calculated by the Army G-8, although the distributions vary from year to year. Army RDTE, which also benefits all components, accounted for about 5 percent; several nonoperational Army responsibilities accounted for about 8 percent. These included civil works (Corps of Engineers), the Joint Improvised Explosive Device Defeat Organization, support to foreign militaries, chemical weapons demilitarization, and maintenance of military cemeteries, among others. These additional Army responsibilities are predominantly performed by Regular Army military and civilian personnel, but could include small percentages of Reserve and Guard contributions.

For FY 2017 data, in Prime awards data (FPDS-NG) the Treasury Account Codes, necessary to assign Compo (component) and Budget Categories, were optional as of June 2016. This affects about $1.5 billion, which is approximately 2.2 percent of the Army dollars obligated in the continental United States in FY 2017. In the Sub awards data (FSRS), Treasury Account Codes are no longer included in the data at all. Therefore, the distribution of spending and indirect effects between components (i.e., Regular Army, Army National Guard, and U.S. Army Reserve) are estimated.[3] To derive these estimates, we determined the average proportional shares of each component for each state during FY 2014 through FY 2016 and applied these portions to Army procurement spending within the state in FY 2017.

---

[3]   This affects only the state-level tables. District-level tables do not break down spending by component, except for the personnel figures, which are from a different source and not affected by the changes in FPDS and FSRS data.

State and District Data

This separate appendix is an ancillary two-volume set available online. It provides detailed results of the analysis, organized by state and congressional district. It includes descriptions of the overall economic effects for each state, then delves into more detail by fiscal year, from 2014 through 2017, concluding with a parsing of the data by congressional district, and providing maps and calculations.

References

Bess, Rebecca, and Zoë O. Ambargis, "Input-Output Models for Impact Analysis: Suggestions for Practitioners Using RIMS II Multipliers," paper presented at the 50th Southern Regional Science Association Conference, New Orleans, La., March 23–27, 2011. As of February 12, 2014:
http://www.bea.gov/papers/pdf/WP_IOMIA_RIMSII_020612.pdf

Coughlin, Cletus C., and Thomas B. Mandelbaum, "A Consumer's Guide to Regional Economic Multipliers," *Federal Reserve Bank of St. Louis Review*, Vol. 73, No. 1, January–February 1991, pp. 19–32.

Defense Manpower Data Center, "Welcome to DMDC," webpage, undated. As of March 2, 2015:
https://www.dmdc.osd.mil/appj/dwp/index.jsp

Deitrick, Sabina, Christopher Briem, Colleen Cain, and Erik R. Pages, *A Comprehensive Assessment of Pennsylvania's Military Installations*, Pittsburgh, Pa.: University of Pittsburgh Center for Social and Urban Research, June 2018.

DMDC—*See* Defense Manpower Data Center.

DoD—*See* U.S. Department of Defense.

Environmental Systems Research Institute, ArcGIS software, USA Base Map data set, 2013.

———, ArcGIS software, U.S. ZIP Code Points File Geodatabase Feature Class, 2016.

ESRI—*See* Environmental Systems Research Institute.

Grady, Patrick, and R. Andrew Muller, "On the Use and Misuse of Input-Output Based Impact Analysis in Evaluation," *Canadian Journal of Program Evaluation*, Vol. 3, No. 2, 1988, pp. 49–61.

Hughes, David W., "Policy Uses of Economic Multiplier and Impact Analysis," *Choices*, Vol. 18, No. 2, 2nd Quarter, 2003.

IMPLAN Group, homepage, undated-a. As of February 12, 2015:
http://implan.com

———, "IMPLAN Data Sources," webpage, undated-b. As of April 5, 2019:
https://implanhelp.zendesk.com/hc/en-us/articles/115009674448-IMPLAN-Data-Sources

———, "Release Notes," webpage, undated-c. As of February 12, 2015:
https://implanhelp.zendesk.com/hc/en-us/articles/115009674368-Release-Notes

Inspector General, U.S. Department of Defense, *Army Needs to Identify Government Purchase Card High-Risk Transactions*, Washington, D.C., DODIG-2012-043, January 20, 2012.

Melcher, David F., and Edgar E. Stanton III, "Army FY 2009 Budget Overview," U.S Army briefing, February 2008.

Office of the Under Secretary of Defense for Acquisition, Technology and Logistics, "Federal Subaward Reporting System (FSRS)," webpage, undated. As of April 24, 2015:
http://www.acq.osd.mil/dpap/pdi/eb/federal_subaward_reporting_system.html

Schnaubelt, Christopher M., Craig A. Bond, Frank Camm, Joshua Klimas, Beth E. Lachman, Laurie L. McDonald, Judith D. Mele, Paul Ng, Meagan Smith, Cole Sutera, and Christopher Skeels, *The Army's Local Economic Effects*, Santa Monica, Calif.: RAND Corporation, RR-1119-A, 2015a. As of July 2018:
https://www.rand.org/pubs/research_reports/RR1119.html

———, *The Army's Local Economic Effects: Appendix B, Volume I: Alabama Through Minnesota*, Santa Monica, Calif.: RAND Corporation, RR-1119/1-A, 2015b. As of May 2015:
http://www.rand.org/pubs/research_reports/RR1119z1.html

———, *The Army's Local Economic Effects: Appendix B, Volume II: Mississippi Through Wyoming*, Santa Monica, Calif.: RAND Corporation, RR-1119/2-A, 2015c. As of May 2015:
http://www.rand.org/pubs/research_reports/RR1119z2.html

Schnaubelt, Christopher M., Craig A. Bond, Cole Sutera, Anthony Lawrence, Judith D. Mele, Joshua Mendelsohn, Christina Panis, and Meagan L. Smith, *The Army's Local Economic Effects*, 2nd. ed., *Appendix B, Vol. I, Alabama Through Minnesota*, Santa Monica, Calif.: RAND Corporation, RR-1119/1-1-A, 2021a. As of March 2021:
http://www.rand.org/pubs/research_reports/RR1119z1-1.html

———, *The Army's Local Economic Effects*, 2nd. ed., *Appendix B, Vol. II, Mississippi Through Wyoming*, Santa Monica, Calif.: RAND Corporation, RR-1119/2-1-A, 2021b. As of March 2021:
http://www.rand.org/pubs/research_reports/RR1119z2-1.html

U.S. Army Environmental Command, *Supplemental Programmatic Environmental Assessment for Army 2020 Force Structure Realignment*, Washington, D.C., June 2014. As of February 27, 2015:
https://aec.army.mil/application/files/6114/9519/9754/Army2020SPEA-1.pdf

U.S. Bureau of Economic Analysis, "Table 1.1.9. Implicit Price Deflators for Gross Domestic Product," National Income and Product Accounts Tables, webpage, March 27, 2015. As of April 10, 2015 (URL no longer works):
http://www.bea.gov/iTable/iTable.cfm?reqid=9&step=3&isuri=1&903=13#reqid=9&step=3&isuri=1&903=13

U.S. Census Bureau, "American Community Survey," undated-a. As of April 2, 2019:
http://www.census.gov/acs/www/

———, "Community Facts," webpage, undated-b. As of May 5, 2015:
http://factfinder.census.gov/faces/nav/jsf/pages/index.xhtml

———, "My Congressional District," webpage, undated-c.

———, "USA Counties," webpage, undated-d. As of May 5, 2015 (URL no longer works):
http://censtats.census.gov/usa/usainfo.shtml

———, "Cartographic Boundary Shapefiles—Congressional Districts," October 2014.

———, "TIGER/Line Shapefiles," April 2015. As of May 4, 2015 (URL no longer works):
http://www.census.gov/geo/maps-data/data/tiger-line.html

U.S. Department of Defense, *Military Compensation*, Greenbooks, undated. As of November 11, 2015:
http://militarypay.defense.gov/References/Greenbooks.aspx

U.S. Department of Defense, Office of the Actuary, *Statistical Report on the Military Retirement System: Fiscal Year 2014*, Alexandria, Va., June 2015. As of April 4, 2019:
https://actuary.defense.gov/Portals/15/Documents/MRS_StatRpt_2014.pdf

———, *Statistical Report on the Military Retirement System: Fiscal Year 2015*, Alexandria, Va., July 2016. As of April 4, 2019:
https://actuary.defense.gov/Portals/15/Documents/MRS_StatRpt_2015%20Final%20v2.pdf?ver=2016-07-26-162207-987

———, *Statistical Report on the Military Retirement System: Fiscal Year 2016*, Alexandria, Va., July 2017. As of April 4, 2019:
https://actuary.defense.gov/Portals/15/Documents/MRS_StatRpt_2016%20v4%20FINAL.pdf?ver=2017-07-31-104724-430

U.S. Postal Service, *Postal Facts 2014*, p. 19.

U.S. Treasury, *Federal Account Symbols and Titles (FAST) Book: Agency Identifier Codes*, Washington, D.C., December 2018. As of April 4, 2019:
https://fiscal.treasury.gov/files/fast-book/fastbook-december-2018.pdf